Push It or Pull It?

written by Rozanne Lanczak Williams
illustrated by Sue Miller

Requests for permission to make copies of any part of the work should be mailed to the following address: School Permissions, Harcourt, Inc., 6277 Sea Harbor Drive, Orlando, Florida 32887-6777.

HARCOURT and the Harcourt Logo are trademarks of Harcourt, Inc.

Printed in China

ISBN-13: 978-0-15-365338-4

ISBN-10: 0-15-365338-8

11   0940   16 15
45004519978

Harcourt
SCHOOL PUBLISHERS

Visit *The Learning Site!*
www.harcourtschool.com

How do you make things go?

You add a force–

A push or a pull.

That is what makes things go!

Push it or pull it.

Push it or pull it.

How can I make it go?

I will push the swing
way up in the air.
When I push it,
it will go!

Push it or pull it.

Push it or pull it.

How can I make it go?

I will push my wheelbarrow
filled with soil.
When I push it,
it will go!

Push it or pull it.

Push it or pull it.

How can we make it go?

We will push and pull.

We will pull and push.

Both will make it go!